Felipe Jose Salomao de A. Melo
Ângelo Just Costa e Silva
Mibson Michel S. Ramos

Evaluation of the adhesion of steel bars to air-entrained concrete

Felipe Jose Salomao de A. Melo
Ângelo Just Costa e Silva
Mibson Michel S. Ramos

Evaluation of the adhesion of steel bars to air-entrained concrete

Study of the adhesion interaction between concrete and steel, through experiment and bibliographical studies

ScienciaScripts

Imprint

Cover image: www.ingimage.com

This book is a translation from the original published under ISBN 978-613-9-69911-7.

Publisher:
Sciencia Scripts
is a trademark of
Dodo Books Indian Ocean Ltd. and OmniScriptum S.R.L publishing group

120 High Road, East Finchley, London, N2 9ED, United Kingdom
Str. Armeneasca 28/1, office 1, Chisinau MD-2012, Republic of Moldova, Europe
Printed at: see last page
ISBN: 978-620-8-14216-2

To my family and friends
And to the Lord Jesus Christ.

Thank you

To God, the source of my greatest inspiration in the face of problems and achievements.

To my family and friends for their support, complicity and understanding.

To the professors of the Civil Engineering Department of the Polytechnic School for their patience over the years of professional conviviality, exchange of knowledge and experiences, which were indispensable for the realisation of this work.

To my colleagues at the Gerdau Group for always providing me with a routine full of new challenges and constant learning.

Summary

The aim of this work is to study the extent to which the addition of an air-entraining agent to concrete can influence adhesion to steel, so that this relationship can be established as one of the parameters when defining or sizing the reinforced concrete to be used. The adherence between steel and concrete is fundamental to the existence of reinforced concrete structures, since the two materials act together to absorb the stresses.

The phenomenon of adhesion is considered complex in terms of the various factors that influence it. It is important to define the force required to release/break the steel bar from the reinforced concrete block. To this end, a concrete mix was cast in reinforced specimens without and with an air admixture and these were subjected to tensile and compressive tests.

It is to be expected that in the tensile test, a greater amount of force was required to pull the steel bar out of the specimen without the air incorporator. And the tensile force required by the machine with the air-incorporated specimen was much lower than the conventional one, concluding that there is an interference in the material's adherence.

Key words: Adhesion, Concrete and traction.

Suoároo

CHAPTER 1

Introduction

The constant evolution of materials and technologies has brought many different materials to the engineering field, as well as various additives and formulas that can alter previously defined values. Adhesion is the property that prevents a bar from slipping in relation to the concrete that surrounds it. It is therefore responsible for the solidarity between steel and concrete, making these two materials work together (PINHEIRO; MUZARDO, 2003).

The forces between concrete and steel are so compatible that the deformation between them is fundamental to the existence of reinforced concrete. This is only possible due to adhesion.

The behaviour of the bond between the reinforcing bar and the surrounding concrete is of decisive importance for the load capacity of reinforced concrete structures. The main characteristics that influence the bond are the type of concrete used (aggregate geometry, additions such as materials with pozzolanic properties or fibres), the geometry of the ribs of the bars and the loading parameters. (TORRE-CASANOVA et al., 2012) The transfer of force from a bar to the concrete is a highly complex phenomenon. There are several factors that influence the behaviour of adhesion, which makes this phenomenon complex and therefore justifies the need to develop research on the subject.

Book Organisation

This book is divided into 8 chapters. Chapter 2 presents the research methodology used. Chapter 3 discusses the concept and theory of steel, Chapter 4 discusses concrete (additives - air-entraining) and Chapter 5 deals with the adhesion of concrete and steel. These chapters present a bibliographical review of the concepts needed to develop the work. Chapter 6 presents the experimental work and the execution of the experiments. Chapter 7 reports the results and discussions of the experiments in the previous chapter. Finally, Chapter 8 presents the conclusions of the work.

1.2 Objectives

1.1.1 General Objective

The general aim of this project is to present an experimental study on the adhesion of steel bars to conventional concrete and concrete with air-entraining admixtures.

1.1.2 Specific objectives

In order to achieve the general objective of this project, the following specific objectives were met:

- Learning about adherence
- Characterise the type of developer
- Observe the behaviour of air-entrained concrete.

- Analyse the results obtained.

Justification

The transfer of force from a bar to concrete is a highly complex phenomenon. The need arose for an in-depth study of the behaviour of adhesion in lightweight and cellular concretes, stimulated by the evolution of materials.

Considering the structural importance of the adherence of steel to concrete, further study is required with a broad view of the various factors that influence adherence.

Understand the major behaviour of the application of air-entraining admixtures to concrete, examining adhesion and the different external and internal factors.

CHAPTER 2

Methodology

In order to carry out the project, an experiment was carried out to study the adherence of the material using the pull-out test (POT - Adapted). The data obtained in previous experiments and the standard governing the adhesion of materials without air-entraining additives (ABNT-NBR 6118, 2004) were analysed.

Initially, we carried out an in-depth study of the various concepts and functioning of the air-incorporating material, after which we provided the materials needed to carry out the experiment.

An experiment was then carried out on pull-out traction, in which the specimens were placed under tensile stress to check the load required for this action. The aim was to understand the influence of the air-entraining agent on adhesion, interfering with load-bearing strength.

CHAPTER 3

Steel

3.1 Definition

According to Chiaverini (2005):

> Steel is an alloy of a relatively complex nature and its definition is not simple since, strictly speaking, commercial steels are not binary alloys. In fact, although its main alloying elements are iron and carbon, they always contain other secondary elements, present due to manufacturing processes. Under these conditions, we can define steel as an Iron-Carbon alloy, generally containing from 0.008% to approximately 2.11% carbon, plus certain secondary elements (such as Silicon, Manganese, Phosphorus and Sulphur), present due to manufacturing processes.

3.1.1 Manufacturing Process

There are various types of raw materials available for making steel. However, due to its lower cost, greater availability and because it is recyclable, which strengthens the aspects of sustainability and socio-environmental responsibility, the basic raw material for the production of steel bars and wires for concrete reinforcement is scrap steel.(ABNT-NBR 7480, 2007)

This rigorously selected scrap is made up of sheet metal scraps, machining chips, used car bodywork, iron parts from disused equipment, among others.

The use of scrap generates a final product with better performance in civil construction. The residual chemical elements usually present in higher percentages in scrap, such as nickel, chrome and tin, result in materials with better mechanical characteristics when compared to steels made from iron ore.

Other raw materials used during the process are:

- Pig iron: a steelmaking product obtained by reducing iron ore, its function is to add carbon, iron and silicon to the product. Carbon and silicon are important sources of energy for the process, through their oxidation generated after the blowing of oxygen. (MITTAL, 2017) Pig iron is the main raw material for steel production in the blast furnace, where a mixture of coke and sinter is also added. It is at this stage that the greatest energy expenditure in the steel industry takes place.
- Ferroalloys (ferro-manganese, ferro-silicon-manganese, ferro-silicon, etc.): used to adjust the chemical composition of steel and give it the necessary mechanical characteristics, such as hardness, ductility and brittleness. Iron-magnesium alloys, for example, are responsible for the addition of phosphorus to steel production, giving

increased hardness, reduced ductility, cold brittleness and increased hot brittleness (PEDRINI; CATEN, 2009).

- Lime: acts as a scorifier, retaining impurities in the metal and forming slag, also acting in the furnace against chemical attacks. Lime provides desulphurisation, which is considered a fundamental stage in steelmaking, since the sulphur present in the production process is detrimental to its properties. Lime can also be replaced by other materials such as marble waste or calcium aluminate in the desulphurisation of steel (COLETI et aL, 2015).
- Oxygen: used to reduce the carbon content of the steel and shorten the melting time, and is a source of heat for the process. The higher the concentration of oxygen in the blowing air, the higher the flame temperature; however, enriching the blowing air with oxygen increases the cost of production. (ASSUNÇÃO, 2011) Throughout the world, the process of manufacturing steel for civil construction is similar. This similarity refers to:
- Manufacturing processes;
- Raw materials used (scrap, pig iron and others);
- Steelmaking, rolling and wire drawing equipment.

The fundamental processes for steel production can be summarised in Figure 2, which looks from the beginning of the extraction of the mineral to its various uses. Each process is detailed below.

3.2.1 - Steel bars for construction: types, standardisation.

3.2.1.1 Types of Steel Bars for Construction

CA-50 Rebar Produced in strict accordance with the specifications of standard NBR 7480, CA-50 Rebar is manufactured with a ribbed surface. It is sold in straight and bent bars with a length of 12 metres and can be found in gauges from 6.3 mm to 32 mm.

Nominal diameter (mm)	Nominal mass] (kg m)	Characteristic yield strength (fy)(MPa)	Strength Limit (MPa)	Stretch, minimum 10 0	Pin diameter for 180° bending (mm)
6,3	0,25	500	1.10 fy	8%	4xDN
8,0	0.40	500	1.10 fy	8°o	4xDN
10.0	0.62	500	1.10 fy	8%	4 x DN
12,5	0.96	500	1.10 fy	8%	4xDN
16.0	1,58	500	1.10 fy	8%	4 x DN
20.0	2.47	500	1.10 fy	8%	6xDN
25,0	3.85	500	1.10 fy	8%	6 x DN
32.0	6,31	500	1.10 fy	8%	8xDN

Table 01- CA50 Rebar Technical Information
Source: (GERDAU, 2009)

CA-25 rebar - Used in reinforced concrete structures, CA25 rebar is also produced in strict accordance with the specifications of standard NBR 7480. The rebar has a smooth surface and is sold in straight or bent bars with a length of 12 metres and can be found in gauges from 6.3mm to 20.0mm.

Nominal diameter (mm)	Nominal mass (kg m)	Characteristic yield strength (fy)(MPa)	Strength Limit (MPa)	Stretch, minimum 10 0	Pin diameter for 180° bending (mm)
6.3	0,25	250	1.20 fy	18%	2 x DN
8.0	0.40	250	1.20 fy	18%	2xDN
10.0	0.62	250	1.20 fy	18%	2 x DN
12,5	0.96	250	1.20 fy	18%	2xDN
16.0	1,58	250	1.20 fy	18%	4 x DN
20.0	2,47	250	1.20 fy	18%	4 x DN

Table 02 - CA2 Rebar Technical Information

Source: (GERDAU, 2009)

CA-60 rebar - Produced in accordance with standard 7480, CA60 is known for its high strength, providing lighter reinforced concrete structures. It is sold in rolls weighing approximately 170 kg, 12 metre long bars, straight or bent, and in stockpiles for industrial customers.

Nominal diameter (mm)	Nominal mass (kgm)	Characteristic yield strength (fy)(MPa)	Endurance Limit (fst)	Stretch. in 10 0	fct/fy ratio	Pin diameter for 180° bending (mm)
4,2	0.11	600	660	5%	>=1,05	5xDN
5,0	0.15	600	660	5%	>=1.05	5xDN
6,0	0,22	600	660	5%	>=1,05	5xDN
7,0	0.30	600	660	5%	>=1,05	5xDN
8,0	0.40	600	660	5%	>=1,05	5xDN
9,5	0.56	600	660	5%	>=1,05	5xDN

Table 03 - CA 60 Technical Information

Source: (GERDAU, 2009)

3.2.2 - Functions of steel in reinforced concrete

Reinforced concrete means concrete with steel bars immersed in it - concrete is considered "reinforced" with steel reinforcement. Reinforced concrete is therefore a composite construction material in which the bond between the concrete and the steel reinforcement is due to cement adhesion and mechanical effects.

The reinforcement bars must absorb the tensile stresses that arise in parts subjected to bending or tension, since concrete has high compressive strength but low tensile strength. Due to adhesion, the deformations of the steel bars and the concrete surrounding them must be equal.

Since fractured concrete cannot keep up with the large deformations of the steel, the concrete cracks in the tensile zone; the tensile stresses must then be absorbed by the steel alone. A simple concrete beam would break abruptly after the first crack, once the low tensile strength of the concrete has been reached, without taking advantage of its high compressive strength.

The reinforcement should therefore be placed in the tensile zone of the structural parts and, whenever possible, in the direction of the internal tensile forces. The high compressive strength of concrete can thus be utilised in bending, in beams and slabs.

CHAPTER 4

Concrete

4.1 Definition

Concrete is a composite material made up of cement, water, fine aggregate (sand) and coarse aggregate (stone or gravel), and air. It can also contain additions (fly ash, pozzolan, active silica, etc.) and chemical additives to improve or modify its basic properties (BASTOS, 2006).

Schematically, paste is cement mixed with water, mortar is paste mixed with sand, and concrete is mortar mixed with stone or gravel, also called simple concrete (concrete without reinforcement).

Lightweight concretes are characterised by a reduction in specific mass compared to conventional concretes. He goes on to say that they can be classified as concrete with light aggregates, cellular concrete and concrete without fines (ROSSIGNOLO, 2009).

4.2 Concrete composition

The first materials to be used in construction were natural stone and wood, with iron and steel being used centuries later. Reinforced concrete only appeared more recently, around 1850.

For a building material to be considered good, it must have two basic characteristics: strength and durability. Natural stone has very high compressive strength and durability, but low tensile strength.

Wood has reasonable strength but limited durability. Steel has high resistance, but requires protection against corrosion.

Reinforced concrete may have arisen from the need to combine the qualities of stone (resistance to compression and durability) with those of steel (mechanical strength), with the advantages of being able to take any shape quickly and easily and providing the necessary protection for steel against corrosion.

Concrete is a composite material made up of cement, water, fine aggregate (sand) and coarse aggregate (stone or gravel), and air. It can also contain additions (fly ash, pozzolan, active silica, etc.) and chemical additives to improve or modify its basic properties, as shown in figure 01 (BASTOS, 2006).

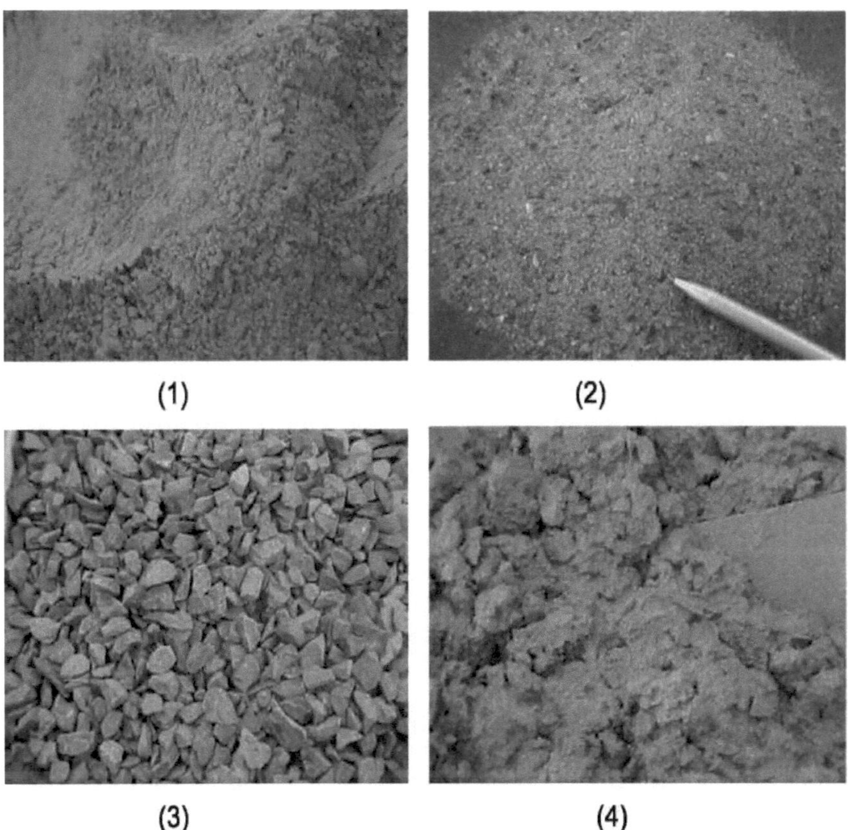

Figure 01 - Concrete Materials - (1) Cement, (2) Fine Aggregate, (3) Coarse Aggregate, (4) Simple Mix with Water - Concrete Source: Author

4.3 Additives

Concrete admixtures are products used in the production of concrete and cement mortars to modify certain properties of the fresh or hardened material. The admixture is often considered the fourth element of concrete if we consider the following sequence: cement, aggregates, water and admixtures.

These are products added during the concrete preparation process, in a quantity no greater than 5% of the mass of cementitious material contained in the concrete, with the aim of modifying concrete properties in the fresh state and\or in the hardened state, except inorganic pigments for the preparation of coloured concrete. (ABNT-NBR 11768, 2011)

The purpose of admixtures varies greatly depending on the application for which the concrete is to be used, so they can increase workability or plasticity, reduce cement consumption, alter, accelerate or delay setting time, reduce shrinkage and increase durability (by inhibiting reinforcement corrosion/neutralising alkali-aggregate reactions).

4.4 - Types of Additives

As described in the standard, the classification of the types of additives found in the chemical industry at the service of the construction industry is as follows: (ABNT-NBR 11768, 2011)

- Plasticising additive (type P): A product that increases the consistency index of concrete while maintaining the amount of mixing water, or that makes it possible to reduce the amount of mixing water by at least 6% in order to produce concrete with a given consistency.

Retarding admixture (type R): Product that increases the setting and setting times of concrete.

Accelerating admixture (type A): A product that reduces the setting and setting times of concrete and accelerates the development of its initial strength.

Plasticising retarding additive (type PR): Product that combines the effects of plasticising and retarding additives.

Plasticising accelerator additive (type PA): Product that combines the effects of plasticising and accelerating additives.

Air-entraining admixture (type IAR): Product that incorporates small air bubbles into the concrete.

Superplasticising admixture (type SP): Product that increases the consistency index of concrete while maintaining the amount of mixing water, or that makes it possible to reduce the amount of mixing water by at least 12 per cent in order to produce concrete with a given consistency.

Superplasticising retarding additive (type SPR): Product that combines the effects of superplasticising and retarding additives.

Accelerating superplasticising additive (type SPA): Product that combines the effects of superplasticising and accelerating additives.

4.5 - Air Incorporator

Air-entraining admixtures reduce the surface tension of water and incorporate air into the concrete, trapping it in millimetre-sized bubbles (0.1 to 0.8 mm). The inclusion of air is preventative and increases the concrete's resistance to detoriation due to freeze-thaw cycles. They create air bubbles that are very stable, strong, small and close together.

Figure 02 - Microscopic photo of the action of the air incorporator.

Source: Narciso Gonçalves da Silva, Giovana Collodetti, Douglas Z. C. M. Pichetti, Philippe Jean Paul Gleize, UTFPR.

The purpose of air-entraining admixtures in fresh concrete is to increase plasticity by reducing friction between solids, reducing permeability and increasing durability (lower w/c), ease of pouring, greater cohesion and less exudation.

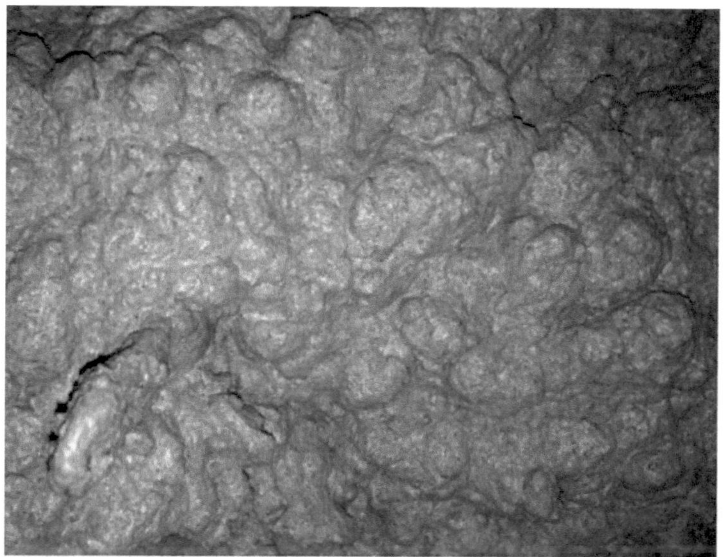
Figure 03 - Photo of concrete with added air.

The effects on hardened concrete increase the concrete's resistance to freeze-thaw phenomena and the obstruction of capillary poles, thus increasing durability.

CHAPTER 5

Adherence

5.1 Adherence

It is the property that prevents a bar from slipping in relation to the concrete that surrounds it, responsible for the solidarity between steel and concrete, making these two materials work together.

The transfer of forces between steel and concrete and the compatibility of deformations between them are fundamental to the existence of reinforced concrete. This is only possible because of adhesion. Anchorage is the fixing of the bar in the concrete so that it can be interrupted.(ABNT-NBR 6118, 2004)

The phenomenon of adhesion involves two aspects: the mechanism of force transfer from the steel bar to the adjacent concrete and the ability of the concrete to resist this force. The transfer of force is made possible by chemical actions (adhesion), friction and mechanical actions, and occurs at different stages of loading and depending on the surface texture of the steel bar and the quality of the concrete.

When anchoring by adhesion, a sufficient length must be provided so that the stress of the bar (tensile or compressive) is transferred to the concrete.

This is known as the anchorage length. In addition, in parts in which, due to constructive provisions or their length, it is necessary to make splices in the bars, a sufficient length must also be guaranteed so that the stresses are transferred from one bar to another in the area of the splice. This is also possible thanks to the adhesion between the steel and the concrete (PINHEIRO; MUZARDO, 2003).

5.2 Types

The types of adhesion are divided into three parts: adhesion, mechanical adhesion and friction. Each one provides different ways of bonding materials.

5.2.1 - Adherence by membership

Adhesion by physical-chemical bonding is characterised by a resistance to the separation of the two materials. It occurs due to chemical bonds at the interface of the bars with the paste, generated during the setting reactions of the cement. For small relative displacements between the bar and the concrete mass surrounding it, this bond is destroyed (FUSCO, 2002).

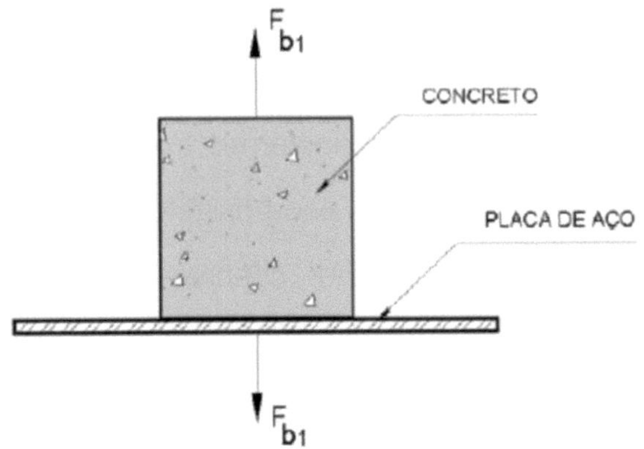

Figure 04- Adherence by Adhesion

Source: (FUSCO, 2002)

5.2. 2- Frictional Adhesion

This portion of adhesion is due to the frictional forces between the steel bar and the concrete, depending on the coefficient of friction given as a function of the surface roughness of the bar and the transverse compression applied by the concrete to the bar.

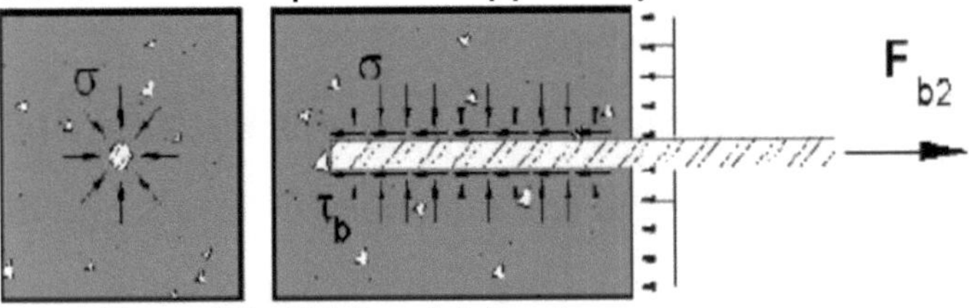

Figure 05 - Frictional Adhesion

Source: (FUSCO, 2002)

5.2. 3- Mechanical adhesion

The portion of adhesion resulting from the formation of concrete consoles between the ribs of the ribbed steel bars. The consoles prevent rapid sliding within the concrete, making this type of adhesion the most effective of the three.

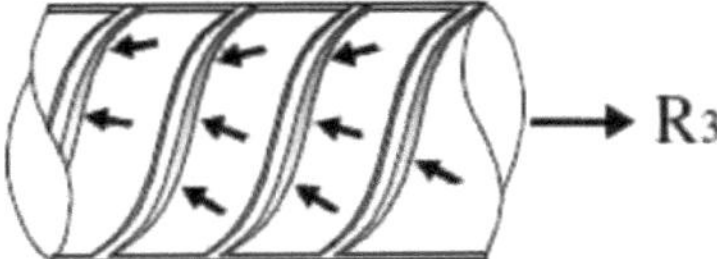

Figure 06 - Mechanical Adhesion Source: (FUSCO, 2002)

5.3 Factors Influencing Adherence

Bond stress can be defined as the stress acting on the bar and the surface of the bar adhering to the concrete. However, there are several factors that can affect its quantification and influence the behaviour of adhesion. Some of these factors are listed below:

- Mechanical strength of concrete;
- Steel yield strength;
- Diameter of the steel bars;
- Spacing between steel bars;
- Position of the bars during concreting;
- Composition of fresh concrete;
- Load age;

- Pressing fresh concrete;
- Surface treatment of steel;
- Concrete cover around the steel bars;
- Shapes and dimensions of steel bar ribs;
- Anchor length;
- Type of load; among others.

5.4 Mechanics of Adhesion

The strength of adhesion is determined by means of different experimental tests, the most common of which is the pull-out of a steel bar inserted into a volume of concrete. The specimens used in pull-out tests determine the global average bond strength, a value that is sufficient to meet the basic design requirements (ABNT-NBR 6118, 2004). The schematic diagram bond stress x relative displacement or slip for a bar with protrusions, determined in a pull-out test.

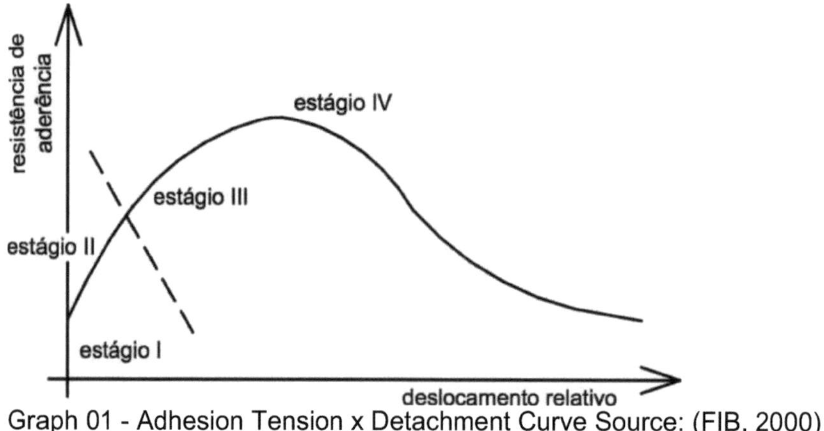

Graph 01 - Adhesion Tension x Detachment Curve Source: (FIB, 2000)

Stage I corresponds to adhesion by adhesion, where rupture occurs with a very small relative displacement, implying that adhesion contributes only a small amount to total adhesion strength.

In stage II, the transfer force is distributed from the bar to the adjacent concrete by the action of the protrusions, which cause the formation of cone-shaped cracks that start at the top of the protrusions. At this stage the relative displacements are still small, caused by the crushing of the concrete under the direct action of the protrusions. The forces on the protrusions are inclined in relation to the bar axis, and can be broken down into directions parallel and perpendicular to the bar axis. The sum of the parallel components equals the adhesion force, and the perpendicular component introduces circumferential tensile stresses around the bar, which can result in radial longitudinal cracks (FIB, 2000).

Stage III begins with the appearance of the first radial crack, and is also maintained by the action of the protrusions on the concrete.

In stage IV, two modes of rupture can occur. If there are no confining stresses in the bar or if they are of low intensity, radial cracks propagate along the entire length of the concrete cover, and rupture occurs due to the cracking action of the concrete.

When the confining stresses are great enough to prevent the concrete cover from cracking, the bond breaks by pulling the bar out of the concrete, changing the mechanism of force transfer from the support of the protrusions in the concrete to friction forces, depending on the shear strength of the existing concrete consoles.(PINHEIRO; MUZARDO, 2003)

5.5 Moments of good or bad grip

Experimental tests have shown that the bond strength of bars positioned in a vertical direction is significantly higher than the bond strength of bars positioned horizontally. For horizontal bars, the distance from the bottom or top of the formwork determines the quality of the adhesion to the concrete.

This is because, during the densification and hardening of the concrete, the sedimentation of the cement and, above all, the phenomenon of exudation, make the concrete in the top layer of the formwork more porous and can reduce the adhesion to half that of the vertical bars.(ABNT-NBR 6118, 2004)

In certain situations, which basically depend on the inclination and position of the reinforcement bar in the concrete mass, the Standard defines situations called "good" and "bad" adhesion.(ABNT- NBR6118, 2004)

The sections of the bars that are in one of the following positions are considered to be in good adhesion:

a) with an inclination greater than 45° to the horizontal;

b) horizontal or inclined at less than 45° to the horizontal, provided that:

- for structural elements with h < 60 cm, located a maximum of 30 cm above the bottom face of the element or the nearest concrete joint.

- for structural elements with h > 60 cm, located at least 30 cm below the top face of the element or the nearest concrete joint.

CHAPTER 6

Experimental Planning

6.1 Experiment

In this work we will observe the behaviour of concrete adhesion to 6.3mm CA50 steel bar by varying only the use of the admixture in the concrete mix. In a clear and objective way, we will study how much the air-entraining agent can influence adhesion with the experiment.

To carry out this experiment, ten wooden moulds with dimensions of 9.5 x 19.5 x 24.5cm were made, nine metres of 10.0mm POP column and a 6.3mm CA50 rebar were purchased, as well as cement, sand and gravel that were already available in UNICAP's materials laboratory.

It was necessary to cut the POP column into ten 22 cm pieces according to the dimensions of the wooden mould, and also to cut the rebar into 40 25 cm pieces to insert them into the mould, one on each side.

From then on, with the exception of the materials for making the concrete, all the materials were ready for use. The next step was to separate the cement, sand and gravel for dosing the concrete mix, which was: 0.55.

Figure 07 - Separation of shapes
Figure 08 - Steel separation
Source: Author

Figure 09-Material assembled Family 01/02
Figure 10 - Assembled material Family 03
Source: Author

6.2 Dosage

The dosage adopted was the same for the entire experiment, only changing with the addition of air.

Table 04 - Traits used

DRAFT - STANDARD				
CEMENT	**SAND**	**BRITAIN**	**WATER/CIMENT**	**Families**
1,00	1,90	2,90	0,55	-
25,00	47,50	72,50	13,75	F I / F II
16	30,4	46,4	8,8	F III

After defining the traits and quantifying the volume of the materials, we went on to manufacture them in order to mould ten test specimens, three of them for family 01 without the air incorporator and with a density of around 2400kg/m^3 considered normal concrete, three of them for family 02 with the addition of 0,05% of additive on top of the cement mass and monitored by time, with a density of around 2000kg/m^3 considered to be lightweight concrete and four for family 03 with the addition of another 0.05% on top of the initial cement mass and monitored by time, with a density of around 1300kg/m considered to be cellular concrete.

We also moulded another 12 cylindrical specimens with the concrete for compressive strength testing. Of the twelve, four were family 01, four were family 02 and four were family 03. CP V-ARI cement was used to make the concrete, which allowed us to carry out the test at an age of seven days. Remember that we are evaluating the time during which the air-entraining agent will be acting on the cement, and directly influencing the adhesion of the steel to the concrete.

Figure 11 - Making the moulds and test specimens.

The tests were carried out at UNICAP's materials laboratory.

6.3 Materials used

The materials used to carry out the work are described below:

- Cement: CP V - ARI
- Medium sand
- Gravel maximum diameter 25mm
- CA-50 rebar - 6.3mm
- POP column - 10.Omm
- Wooden mould - 9,5 x 19,5 x 24,5cm
- Air-entraining additive

6.4 Description of Tests

The tests were divided into two parts.

- The first was the compressive strength test of the cylindrical specimens. This test took place when the concrete was seven days old and in accordance with the standardisation described in NBR 5739 - Compression test of cylindrical specimens.

Figure 12 - Compression test
Figure 13 - Separating the cp's

- The second test involved pulling the rebar out of the reinforced rectangular specimen. For this test we used the material tensile test as a reference. It is important to make it clear that NBR 6152 was only used as a reference, as there is no specific standard for the type of test and how it was conducted.

Figure 14- Pull-out test

Figure 15 - Machine experiment

6.5 Study variables

For cylindrical specimens Exposure conditions:

- Saturated in water from the utility company (Compesa).
- Number of samples with Family I: 04 bodies
- Number of samples with Family II: 04 bodies
- Number of samples with Family III: 04 bodies
- For rectangular reinforced concrete blocks:

- Exposure conditions:

- Dry condition depending on the steel so as not to oxidise it

- Number of samples with Family I: 03 blocks

- Number of samples with Family II: 03 blocks

- Number of samples with Family III: 04 blocks

CHAPTER 7

Results and discussions

7.1 Compression Strength Test

After the tests, it was found that the strength of the specimens made from family I was higher than that of the specimens made from families II and III.

This was to be expected, given that the additive ratio directly influences the strength of the concrete. This variable, time and additive, is the main parameter controlled in the work to find the ideal density.

It was found that in family I, the density obtained is 2550kg/m^3 and there is no addition of the Air incorporator, the specimens had minimal loss of volume after 7 days and showed a fck = 21.56 Mpa, while in family II there was an addition of 0.05% of the initial cement mass where monitoring began.

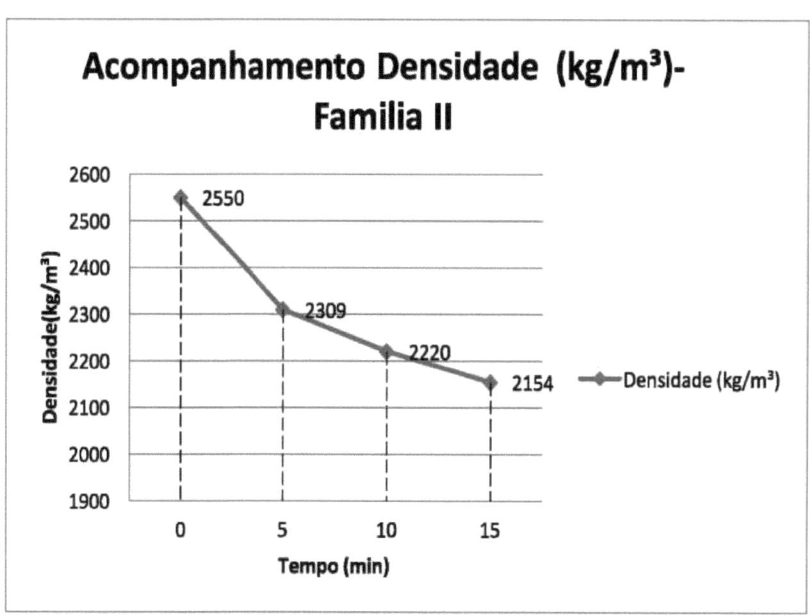

Graph 02 - Monitoring Family Density II

After 15 minutes it was found that the density had reached 2154 kg/m^3 and the material was moulded at this density. The specimens were broken after 7 days, with minimal loss in volume and a fck= 2.7 Mpa. In family III, a further 0.1% of the initial mass of cement was added. After 60 minutes, the density was found to be 1454 kg/m^3 and the material was moulded at this density.

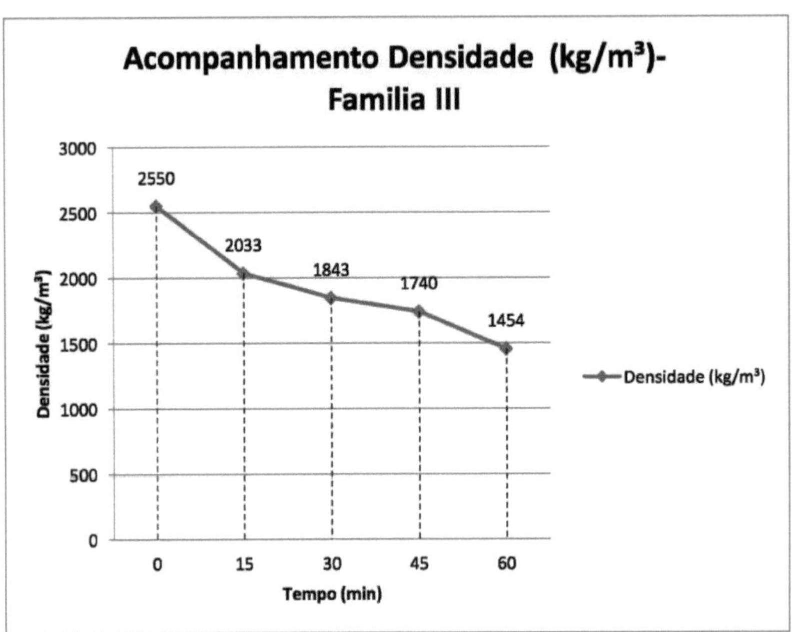

Graph 03 - Monitoring Family Density III

7.2 Pull-out test

A very different behaviour is expected in the various families tested: in family I, where the concrete used was conventional (without additives), adhesion should be normalised; in family II, air-entraining was added and it is expected that there will be less adhesion of the steel to the concrete; in family III, even less adhesion is expected.

CHAPTER 8

Final considerations

Although it was expected that there would be a difference in the force needed to pull the steel out of the concrete, there was no way of determining how much and how this force would vary beforehand.

The results obtained and the results expected in the work show that it is possible to make this prediction more accurately.

However, in order for this to be achieved, it would be necessary to refine this study in order to compare other specimens, with variations in the type of steel used (plain and ribbed), with variations in the amount of air incorporator used in the mix of the mix, being able to use a larger number of specimens for comparison and quantification.

The conclusion is that the adhesion between concrete and steel can be directly influenced by the addition of the air-entraining additive, leaving further study of the chemical bonds and the fragility of these bonds.

This work, therefore, did not obtain satisfactory results

because the equipment supplied did not provide reliable results. An in-depth study of the use of air-entraining in concrete could be carried out with a certain limit so as not to jeopardise the adhesion and function of reinforced concrete.

CHAPTER 9

Bibliographical references

ABNT-NBR 11768, A. B. DE N. T. **Additives for Portland Cement Concrete requirementsBrazil,** 2011.

ABNT-NBR 6118, A. B. DE N. T. **Projecto** de **estruturas de concreto - ProcedimentoPackaging Boston Mass,** 2004.

ABNT-NBR 7480, A. B. DE N. T. **Steel for reinforcement in reinforced concrete structures - SpecificationPackaging Boston** MassBrazil, 2007.

ASSUNÇÃO, C. S. Reducing Oxygen Consumption in a Blast Furnace by Increasing the Energy Efficiency of Air Heaters. **Tecnologia em Metalurgia Materiais e Mineração,** v. 8, n. 2, p. 73-79, 2011.

BASTOS, P. S. D. S. Fundamentals of Reinforced Concrete. In: **Universidade Estadual Paulista,** [s.l: s.n.]. p. 98.

COLETI, J. L. et al. Use of marble residue and calcium aluminate in synthetic steel desulphurising slags. **Technology in Metallurgy, Materials and Mining,** v. 12, n. 3, p. 188-194, 2015.

FIB, F. I. D. B. Structural concrete - Textbook on behaviour, design and performance. In: [s.l: s.n.].

FUSCO, P. B. **Técnicas de Armar as Estruturas de Concreto.** 4ª ed. São Paulo: [s.n.].

GERDAU, S. **Characteristics of Steel.** Available at: <https://www.gerdau.com/br/pt/productsservices/products/Documen t Gallery/Catalogue Civil Construction.pdf>. Accessed on: 10 March 2013.

MITTAL, A. **Manual of the Manufacturing Process of CA50S, CA25 and CA60 Ribbed.**

PEDRINI, D. C.; CATEN, C. S. TEN. **Statistical Modelling for the.**

NATIONAL PRODUCTION ENGINEERING MEETING. **Proceedings...**Salvador: 2009

PINHEIRO, L. M.; MUZARDO, C. D. Adherence and anchoring. In: **Class Notes**. 1. ed. São Paulo: [s.n.].

ROSSIGNOLO, J. A. **Structural lightweight concrete: Production, properties, microstructure and applications.** 1. ed. São Paulo: [s.n.].

TORRE-CASANOVA, A. et al. Confinement effects on the steel-concrete bond strength and pull-out failure. **Engineering Fracture Mechanics,** v. 97, n. 1, p. 92-104, 2012.

Printed by Books on Demand GmbH, Norderstedt / Germany